ICS 11.020

SCM

世界中医药学会联合会
World Federation of Chinese Medicine Societies

SCM 56-2020

中药饮片内包装

Inner Pack of Decoction Pieces of Chinese Medicine

世界中联国际组织标准
International Standard of WFCMS

2020-7-2发布实施
Issued & implemented on July 2, 2020

中医古籍出版社
Publishing House of Ancient Chinese Medical Books

图书在版编目（CIP）数据

中药饮片内包装：汉英对照 / 世界中医药学会联合会著 . -- 北京：中医古籍出版社，2025.7
ISBN 978-7-5152-2451-0

Ⅰ . ①中… Ⅱ . ①世… Ⅲ . ①饮片—包装材料—汉、英 Ⅳ . ① TQ460.6

中国版本图书馆 CIP 数据核字 (2022) 第 011891 号

中药饮片内包装
世界中医药学会联合会　著

责任编辑　张　楚
封面设计　艺点锦秀
出版发行　中医古籍出版社
社　　址　北京市东城区东直门内南小街 16 号（100700）
电　　话　010-64089446（总编室） 010-64002949（发行部）
网　　址　www.zhongyiguji.com.cn
印　　刷　北京市泰锐印刷有限责任公司
开　　本　880mm×1230mm　1/16
印　　张　1.75
字　　数　47 千字
版　　次　2025 年 7 月第 1 版　2025 年 7 月第 1 次印刷
书　　号　ISBN 978-7-5152-2451-0
定　　价　36.00 元

目　　次

前　言

请注意本文件的某些内容可能涉及专利。本文件的发布机构不承担识别专利的责任。

主要起草单位：江西中医药大学，深圳市卫生健康委员会，江西省卫生健康委员会，德国汉堡大学汉萨美安中医中心，苏州信亨自动化科技有限公司，深圳统标科技有限公司，深圳市标准技术研究院，深圳市卫生健康发展研究和数据管理中心，澳洲全国中医药针灸学会联合会，中国中药控股有限公司，韩国庆熙大学，澳大利亚澳华中医学会，北京中医药大学东方医院，美国杏林健康科技有限公司。

参与起草的单位：安徽广印堂中药股份有限公司，澳大利亚中医药学会，中国人民解放军总医院第五医学中心，北京中医药大学，北京中医药大学东直门医院，亳州市永刚饮片厂有限公司，汕头市澄海区中医医院，广州市黄埔区穗东街夏园社区卫生服务站，广州市健桥惠泽有限公司，加拿大中医药学院，江西江中中药饮片有限公司，江西省药品检验检测研究院，美国加州中医药大学，美国金禾中医药有限公司，美国纽约中医学院，南方医科大学深圳医院，汕头大学医学院第二附属医院，汕头市澄海区人民医院，汕头市澄海区溪南镇卫生院，汕头市中心医院，汕头市中医医院，上海中医药大学，深圳市华辉药业有限公司，深圳市罗湖区中医院，深圳市人民医院，深圳市药品检验研究院，深圳市中医药企业标准联盟，深圳市中医药学会，深圳市中医院，首都医科大学附属北京地坛医院，香港浸会大学，中国食品药品检定研究院中药民族药检定所，中国中医科学院中医药信息研究所，中山市中医院。

本文件主要起草人：吴培凯，Thomas Friedemann（德国），廖利平，李海燕，路海英，陈明人，左铮云，杨明，朱卫丰，李锦轩，邓雪枝，姚克勤，林子强（澳大利亚），张翼（澳大利亚），易炳学，兰青山，Yang Woongmo（韩国），张力。

本文件参与起草人及审阅专家（按姓氏拼音排序）：蔡楚喜，蔡沛彪，蔡文伟，蔡晓纯，蔡永江，陈健星，陈镍，陈权韩，谌瑞林，陈玩珊，陈业孟（美国），程仕萍，丛佳林，杜建强，高广印，龚千锋，古旭明，关志宇，郭晓秋，何杜朋，何浪，胡世平，胡新华，华国栋，黄剑帆，黄举凯，黄展辉，简晖，李静，李锐键，李顺民，李咏，梁葆山（美国），梁晖宙（美国），林韩，林树旭，刘宝河，刘炽京，刘荣华，刘希伟，刘勇，罗光明，吕爱平（中国香港），马双成，穆风杨，彭映梅，王伽伯，王淑红，王伟群，肖小河，熊汉鹏，许素文，许妍，徐子渠，薛春苗，杨具荣，杨西晓，杨晓晖，杨银芳，余亚薇，原文鹏，袁晓宁（加拿大），Ron Zaidman（美国），曾长龙，曾庆明，查青林，章德林，张金莲，张萍，张寿文，章文春，赵恒伯，赵永红，赵振萍（美国），郑建华，周平生，邹健生。

本文件的起草程序遵守了世界中医药学会联合会发布的《世界中医药学会联合会国际组织标准管理办法》和 SCM 0001–2009《标准制定和发布工作规范》。

引　言

中药饮片内包装存在包装不规范、不符合卫生要求、不利于药品储存以及包装材料与中药饮片的性质不相适应等问题，严重影响了中药质量和用药的安全性、有效性，制约了中药现代化、国际化进程。针对当前中药饮片存在无包装或包装不符合规定的情况，为规范中药饮片包装编制本文件，旨在建立、健全中医药标准服务体系与认证制度，规范行业行为和职业操守，营造中药种植、生产、流通、用药的安全与有效的良好氛围，促进中药饮片贸易公平、公正和科学发展，让世界各国人民共享中医药发展成果，为全人类的健康作出应有贡献。

本文件与 ISO 18668–1《中药编码系统 – 第 1 部分：中药编码规则》、ISO 18668–2《中药编码系统 – 第 2 部分：中药饮片的编码》、ISO 18668–3《中药编码系统 – 第 3 部分：中药材的编码》、ISO 18668–4《中药编码系统 – 第 4 部分：中药配方颗粒的编码》、ISO 20333《中药在供应链管理中的编码与表示》和 ISO 20334《中药方剂编码系统》等 6 项 ISO 标准，以及 GB/T 31773《中药方剂编码规则及编码》、GB/T 31774《中药编码规则及编码》、GB/T 31775《中药在供应链管理中的编码与表示》等 3 项中国国家标准等药品标准相衔接。

中药饮片内包装

1 范围

本文件规定了中药饮片内包装的材料、外观、规格、标签等要求。
本文件适用于中药饮片内包装的生产、使用和监督管理。

2 规范性引用文件

下列文件中的内容通过文中的规范性引用而构成本文件必不可少的条款。其中，注日期的引用文件，仅该日期对应的版本适用于本文件；不注日期的引用文件，其最新版本（包括所有的修改单）适用于本文件。

ISO 22217 中医药 – 中药材、中药饮片贮藏规范
ISO 18668–1 中医药 – 编码系统 第 1 部分 中药编码规则
ISO 18668–2 中医药 – 编码系统 第 2 部分 中药饮片的编码
ISO 20333 中医药 – 中药在供应链管理中的编码与表示
ASTM D3981 一般用途及包装应用中等密度聚乙烯制成的聚乙烯薄膜
ASTM D4635 通用和包装用低密度聚乙烯薄膜的标准规范

3 术语和定义

下列术语和定义适用于本文件。

3.1
内包装

直接与中药饮片接触的包装。

3.2
中药饮片

药材经过炮制后可直接用于中医临床或制剂生产使用的处方药品。
［来源：ISO 18668–1: 2016，3.3］

3.3
色标

在中药饮片的包装袋或标签上，使用不同的颜色代表不同的规格。

3.4

潘通色卡

国际通用的标准色卡之一，涵盖印刷、纺织、塑胶、绘图、数码科技等领域的色彩沟通系统。

3.5

包材

适用于本文件的各类内包装材料。

4 外观要求

4.1 应符合相关国家和地区药品监督管理等部门对药品标签的要求。

4.2 包装袋应密封完好，表面干净、光洁、无破损、无松弛和异物附着、无脱层及其他异常情况。

4.3 包装表面应平整，无活褶。裁剪应分切整齐，不得带有毛刺。

4.4 连续式内包装相邻 2 个包装间连接部分应透明，透明部分长度不小于 20 mm。

4.5 贵细中药饮片应采取双重包装等有效措施，以免在贮运过程中因装卸碰撞等导致外包装破损，造成污染和损失。

4.6 按照麻醉和毒性药品管理的中药饮片，内包装上需贴有相应的明显警示标志。

5 材料要求

5.1 应符合相关国家和地区对药品（或食品）包装材料的要求，禁止使用含“氯”成分和再生利用的有毒材料，应从符合资质的生产企业采购。

5.2 应选用与中药饮片性质相适应的包装材料和容器。

5.3 应无毒、无菌、清洁、干燥、无污染、无异味，应透明或部分透明，以便能直接观察内装饮片。

5.4 应具有一定的通透性、防潮性、抗压性和环保性，可以回收利用或可降解。

5.5 内包装材料应一次性使用，不可重复使用。

5.6 按照麻醉和毒性药品管理的中药饮片包装材料应符合当前国家相关法律、法规及标准的规定。

5.7 包装材料的种类和适用范围如下：

5.7.1 聚乙烯塑料薄膜应符合 ASTM D3981 和 ASTM D4635 中的规定，适用于不易霉变、不易虫蛀的中药饮片以及手工定量包装的中药饮片。

5.7.2 包装用聚乙烯吹塑薄膜应符合 ASTM D3981 和 ASTM D4635 中的规定，包括 4 种类型：纸板复合薄膜（纸 / 塑）、纤维复合薄膜（纤维 / 塑）、多层复合薄膜（塑 / 塑）、铝箔复合薄膜（金属 / 塑），适用于易霉变、易被虫蛀的中药饮片。

6 包装剂量规格设定

6.1 设定原则

中药饮片内包装规格设定应遵循“临床常用剂量要求、品规最少及高频多贵”的原则。

6.2 规格要求

6.2.1 内包装中药饮片规格包括13种，分别为：1g、3g、5g、6g、9g、10g、12g、15g、30g、250g、500g、1000g、1000g以上。

6.2.2 内包装中药饮片的装量误差应符合表1要求。

表1 内包装中药饮片装量差异要求

标示质量或标示装量	质量或装量差异限度
1g	±10%
3g、5g	±6%
6g、9g、10g	±5%
12g、15g、30g	±4%
250g、500g、1000g、1000g以上	不少于标示质量或装量的97%

注1：贵细中药饮片可另作特殊要求。

注2：1~30g规格装量差异检查法为取供试品10包，分别称定每包内容物的质量（称取质量应准确至所取质量的1%），每包装量与标示装量相比较，超出装量差异限度要求的不能多于2包，且不能有1包超出差异限度要求1倍。

注3：250g及以上规格装量差异检查法为取供试品3包，分别称定每包内容物的质量（称取质量应准确至所取质量的1%），求出每包装量与平均装量，每包装量与平均装量应符合表1规定。如有1包不符合规定，则另取3包复试。初试或复试，3包装量应全部符合表1规定。

6.3 其他要求

6.3.1 凡具有麻醉功能的药物，如罂粟壳等，其内包装要求严格按照相关国家和地区的规定执行。

6.3.2 按照麻醉和毒性药品管理的中药饮片及《中华人民共和国药典》2020年版注明有毒的中药饮片，其内包装装量规格不能超过相关国家或地区规定的最大剂量。

6.3.3 凡不以重量为剂量单位的中药饮片，如蜈蚣（条）等，可不设定内包装品规，调剂时应按处方标定的剂量，随方处理。

7 标签要求

7.1 中药饮片内包装必须印有或者贴有标签。

7.2 标签中的文字应清晰易辨，表述科学、规范、准确。

7.3 标签中的标识应清楚醒目，不能有印字脱落或者粘贴不牢等现象，不得以粘贴、剪切、涂改等方式进行修改或者补充。

7.4　标签中禁止使用未经注册的商标，禁止使用未经相关国家或地区食品药品管理部门批准的中药饮片名称。

7.5　按照麻醉药品管理的中药饮片和毒性中药饮片的标签，应符合相关国家或地区的有关规定。

7.6　对贮藏有特殊要求的药品，应在标签的醒目位置注明。

7.7　药品包装内不能夹带任何未经批准的介绍或宣传产品、企业的文字以及音像等资料。

7.8　中药饮片的标签应至少包含以下内容：

a）商标；

b）名称（应以 ISO 18668-2 的中药饮片名称为准）；

c）产地；

d）装量规格；

e）条码（含二维码）（依据 ISO 18668-1）；

f）产品批号；

g）生产日期；

h）生产企业；

i）执行的质量标准。

8　二维码要求

8.1　二维码面积

在能够快速识别的前提下，二维码面积应大于包材面积的十六分之一，以不干胶纸加贴标签方式或直接印制于包材外。

8.2　二维码扫描结果

二维码扫描结果应至少包含以下内容：

a）企业名称；

b）商品名称；

c）商品条码；

d）产品种类；

e）中药编码（依据 ISO 18668-1，ISO 18668-2）；

f）品牌；

g）规格 / 型号；

h）产地；

i）贮藏方法（依据 ISO 22217）；

j）净含量；

k）执行标准（《中华人民共和国药典》2020 年版或相关标准）；

l）产品说明（功效）；

m）企业网址；

n）生产厂家；

o）生产地址；

p）备案有效期。

9 色标要求

9.1 色标应醒目、色差大，同一规格不同品种使用同一颜色。

9.2 色标应符合相关国家或地区的有关规定，印刷内容与色泽应满足要求。

9.3 色标应附于包装袋或标签上，标示于包材中部，呈条形色带，色带面积应大于包材面积的四分之一。

9.4 色标采用国际通用的潘通色卡，中药饮片内包装规格与对应色标可参见附录A。

9.5 色标采用以下印制方法：

a）将不同颜色直接印制在内包装上。此法适用于用色带打印标签的内包装和在透明塑料袋上印刷专用标签的内包装。

b）使用多种不同底色的不干胶纸印制标签，每种颜色代表一种规格。此法适用于全透明塑料袋等加贴标签的内包装。

附录 A
（资料性）
中药饮片内包装规格与潘通色卡对应色标

中药饮片内包装规格与潘通色卡的对应色标，如表 A.1 所示。

表 A.1　中药饮片内包装规格与潘通色卡对应色标

潘通色卡编号	颜色名称	RGB	HEX/HTML	实物样品色	对应规格
P 57-8 C	红色	211 22 59	D3163B		1g
16-1356 TCX	橙色	244 115 39	F47327		3g
299 C	蓝色	0 163 224	00 A3E0		5g
20-0084 TPM	石竹色	209 198 185	D1C6B9		6g
1535 C	褐色	148 69 11	94450 B		9g
355 C	薄绿色	0 150 57	009639		10g
2597 U	紫色	130 90 164	825 AA4		12g
14-0760 TCX	黄色	255 212 0	FFD400		15g
8201 C	淡钢蓝色	103 137 156	67899 C		30g

注：根据临床实际情况，30g 以上的规格，不做色标规定。

参考文献

[1]国家药典委员会.中华人民共和国药典(英文版)[M].北京:中国医药科技出版社，2022.

[2]国家药典委员会.中华人民共和国药典(一部)[M].北京:中国医药科技出版社，2022.

[3]国家标准化管理委员会.包装术语：第1部分　基础：GB/T 4122.1—2008[S].北京：中国标准出版社，2011.

[4]全国塑料制品标准化技术委员会.包装用聚乙烯吹塑膜：GB/T 4456—2008[S].北京：中国标准出版社，2011.

[5]国家医药管理局.药品包装用复合膜(通则):YY 0236—1996[S].北京：中国标准出版社，1998.

[6]廖利平，吕爱平，曾庆明.中医药国家标准制定实战录：中医药编码系统国家技术标准解读[M].北京:中国标准出版社，2016.

[7]吴培凯，徐美渠.中药与中药方剂的编码及功效速查手册[M].北京：中国标准出版社，2017.

Foreword

Note that certain contents of this document may involve certain patents. The issuing authority of this document assumes no statutory duty to identify the patent.

The main drafting organizations : Jiangxi University of Chinese Medicine, Health Commission of Shenzhen Municipal, Health Committee of Jiangxi Province, Hanse Merkur Center for Traditional Chinese Medicine at the University Medical Center Hamburg- Eppendorf gGmbH in Germany, Suzhou Xinheng Automation Technology Co., Ltd. Shenzhen Unified Standard Technology Co., Ltd., Shenzhen institute of standards and technology, Shenzhen Health Development Research and Data Management Center, Federation of Chinese Medicine and Acupuncture Societies of Australia (FCMA), China Traditional Chinese Medicine Holdings Co., Ltd., Korea KyungHee University, Australian Austro-Central China Medical Association, Dongfang Hospital of Beijing University of Chinese Medicine and USA Hingli Health and Technology Co.,ltd.

The drafting organizations participants : Anhui Guangyintang Traditional Chinese Medicine Co., Ltd., Chinese Medicine & Acupuncture Society of Australia, The Fifth Medical Center of the General Hospital of the Chinese People's Liberation Army, Beijing University of Chinese Medicine, Dongzhimen Hospital of Beijing University of Chinese Medicine,Bozhou Yonggang Decoction Pieces Factory Co., Ltd., Chenghai Hospital of Traditional Chinese Medicine, Xiayuan Community Health Service Station of Suidong Huangpu Guangzhou,Guangzhou Jianqiao Huize Co., Ltd.,College of Traditional Chinese Medicine of Canada, Jiangxi Jiangzhong TCM Decotion Pieces Co., Ltd., Jiangxi Institute for Drug Control, Five Branches University of California USA, USA Kamwo Meridian Herbs, New York College of Traditional Chinese Medicine, Shenzhen Hospital of Southern Medical University, The Second Affiliated Hospital of Shantou University Medical College,People's Hospital of Chenghai Shantou,Health Center of Xi'nan Chenghai Shantou, Shantou Central Hospital, Shantou Hospital of Traditional Chinese Medicine, Shanghai University of Traditional Chinese Medicine, Shenzhen Huahui Pharmaceutical Co., Ltd., Shenzhen Luohu District Hospital of Traditional Chinese Medicine, Shenzhen People's Hospital, Shenzhen Institute for Drug Control, Shenzhen Standards Alliance for Traditional Chinese Medicine Enterprises, Shenzhen Society of Traditional Chinese Medicine, Shenzhen Hospital of Traditional Chinese Medicine,Beijing Ditan Hospital of Capital Medical University,Hong Kong Baptist University, Inspection Office of Traditional Chinese Medicine of National Institute for Food and Drug Control,Institute of Information on Traditional Chinese Medicine of China Academy of Chinese Medical Sciences and Zhongshan Hospital of Traditional Chinese Medicine.

The main drafters : Wu Peikai, Thomas Friedemann (Germany), Liao Liping, Li Haiyan, Lu Haiying, Chen Mingren, Zuo Zhengyun, Yang Ming, Zhu Weifeng, Li Jinxuan, Deng Xuezhi, Yao Keqin, Lin Ziqiang (Australia), Kevin Chang (Australia), Yi Bingxue, Lan Qingshan, Yang Woongmo (South Korea) and Zhang Li.

Drafting participants and reviewing experts (in alphabetical order of surname): Cai Chuxi, Cai

Peibiao, Cai Wenwei, Cai Xiaochun, Cai Yongjiang, Chen Jianxing, Chen Nie, Chen Quanhan, Chen Ruilin, Chen Wanshan, Chen Yemeng (United States), Cheng Shiping, Cong Jialin, Du Jianqiang, Gao Guangyin, Gong Qianfeng, Gu Xuming, Guan Zhiyu, Guo Xiaoqiu, He Dupeng, He Lang, Hu Shiping, Hu Xinhua, Hua Guodong, Huang Jianfan, Huang Jukai, Huang Zhanhui, Jian Hui, Li Jing, Li Ruijian, Li Shunmin, Li Yong, Liang Baoshan (United States), Liang Huizhou (United States), Lin Han, Lin Shuxu, Liu Baohe, Liu Chijing, Liu Ronghua, Liu Xiwei, Liu Yong, Luo Guangming, Lü Aiping (Hong Kong China), Ma Shuangcheng, Mu Fengyang, Peng Yingmei,Wang Jiabo, Wang Shuhong, Wang Weiqun, Xiao Xiaohe, Xiong Hanpeng, Xu Suwen, Xu Yan, Xu Ziqu, Xue Chunmiao, Yang Jurong, Yang Xixiao,Yang Xiaohui, Yang Yinfang, Yu Yawei, Yuan Wenpeng, Yuan Xiaoning (Canada), Ron Zaidman (United States), Zeng Changlong, Zeng Qingming, Zha Qinglin, Zhang Delin, Zhang Jinlian, Zhang Ping, Zhang Shouwen, Zhang Wenchun, Zhao Hengbo, Zhao Yonghong, Zhao Zhenping (United States), Zheng Jianhua, Zhou Pingsheng and Zou Jiansheng .

The drafting procedure of this document complies with the Standard Management *Measures for International Organizations of the World Federation of Chinese Medicine Societies* and SCM 0001-2009 Standards Development and Publication Work Specification issued by WFCMS.

Introduction

There are some problems in the internal packaging of traditional Chinese medicine decoction pieces, such as non-standard packaging, non-compliance with hygiene requirements, unfavorable storage of traditional Chinese medicine decoction pieces, incompatibility of packaging materials with the nature of traditional Chinese medicine decoction pieces, etc., which seriously affect the quality of traditional Chinese medicine as well as safety and effectiveness of medication, thereby restricting the process of modernization and internationalization of traditional Chinese medicine. In view of the fact that there is no package or the package does not conform to the regulations for the traditional Chinese medicine decoction pieces, this document has been compiled in order to standardize the packaging of traditional Chinese medicine decoction pieces, aiming to establish and perfect the standard service system and certification system of traditional Chinese medicine, standardize the industry behavior and professional ethics, create a safe and effective environment for the cultivation, production, circulation and medication of traditional Chinese medicine, promote a good atmosphere for fair, just, harmonious and scientific development of trade, enable people from all over the world to share the achievements of the development of traditional Chinese medicine, and making due contributions to the health of all mankind.

This document is aligned with six ISO standards, including ISO 18668-1 *Traditional Chinese Medicine-Coding System for Chinese Medicines-Part 1:Coding Rules for Chinese Medicines,* ISO 18668-2 *Traditional Chinese Medicine-Coding System for Chinese Medicines-Part 2:Codes for Decoction Pieces,* ISO 18668-3 *Traditional Chinese Medicine-Coding System for Chinese Medicines-Part 3:Codes for Chinese Materia Medica,* ISO 18668-4 *Traditional Chinese Medicine-Coding System for Chinese Medicines-Part 4:Codes for Granule Forms of Individual Medicinals for Prescriptions,* ISO 20333 *Traditional Chinese Medicine - Coding Rules for Chinese Medicines in Supply Chain Management,* and ISO 20334 *Traditional Chinese Medicine - Coding System of Formulae,* as well as three Chinese National Standards (GB/T) such as GB/T 31773 *Coding Rules for Chinese Medicinal Formulae and Their Codes,* GB/T 31774 *Coding Rules for Chinese Medicines and their Codes, and* GB/T 31775 *Coding Rules for Chinese Medicines in Supply Chain Management.*

Traditional Chinese Medicine – Inner Pack of Decoction Pieces

1 Scope

This document specifies requirements for inner pack of decoction pieces including materials, appearance, specifications,labels and other relevant requirements.

This document is applicable to the manufacturing, use, supervision and management of inner pack for decoction pieces.

2 Normative References

The following referenced documents are indispensable for the application of this document. For dated references, only the edition cited applies. For undated references, the latest edition of the referenced document (including any amendments) applies.

ISO 22217 *Traditional Chinese Medicine - Storage Reqairements for Raw Materials and Decoction Pieces*

ISO 18668-1 *Traditional Chinese Medicine - Coding System for Chinese Medicines - Part 1: Coding Rules for Chinese Medicines*

ISO 18668-2 *Traditional Chinese Medicine - Coding System for Chinese Medicines - Part 2: Codes for Decoction Pieces*

ISO 20333 *Traditional Chinese Medicine - Coding Rules for Chinese Medicines in Supply Chain Management*

ASTM D3981 *Standard Specification for Polyethylene Films Made From Medium-density Polyethylene for General Use and Packaging Applications*

ASTM D4635 *Standard Specification for Plastic Films Made From Low-density Polyethylene and Linear Low-density Polyethylene for General Use and Packaging Applications*

3 Terms and Definitions

The following terms and definitions apply to this document.

3.1
Inner Pack

Package in direct contact with decoction pieces.

3.2

Decoction Pieces

Prescription medicinals processed from Chinese materia medica under the direction of Traditional Chinese Medicine and processing methods for Chinese medicines, which can be directly used in clinical practice or the production of prepared medicines

[SOURCE: ISO 18668-1:2016, 3.3]

3.3

Colour Code

Different colours used on the package or label of decoction pieces for indicating different specifications.

3.4

Pantone Colour Card

A kind of international standard colour card, covering colour communication system of many fields including printing, textile, plastics, drawing and digital technology, etc.

3.5

Packing Materials

Any materials of inner pack applicable to this document.

4 Appearance Requirements

4.1 Drug labelling requirements of departments related to drug supervision and administration shall be followed.

4.2 Bags for inner pack of decoction pieces shall be securely sealed.The surface shall be clean and smooth without damage, relaxation and foreign matter attaching, delamination and other abnormalities.

4.3 The surface of packaging shall be smooth without pleats. And cutting of packaging shall be neat and with no burr.

4.4 For continuous inner packaging, the connection part between two adjacent packages shall be transparent and its length shall be no less than 20mm.

4.5 For precious and scarce decoction pieces, effective measures such as double packaging shall be taken to avoid losses and pollution resulted from damaged packaging in the process of storage and transportation due to loading and unloading collision.

4.6 Decoction pieces which are managed according to narcotic and toxic drugs, shall be packed in a special way, with notable corresponding warning signs attached on inner pack.

5 Material Requirements

5.1 Inner pack materials shall be consistent with national standards of drug (or food) packaging materials and purchased from qualified manufacturing enterprises. Materials containing chlorinated compounds and recycled toxic materials shall be prohibited.

5.2 Inner pack materials and containers shall be selected in accordance with the properties of decoction pieces.

5.3 Inner pack materials shall be non-toxic, sterile, clean, dry,pollution and peculiar smell.And shall be transparent or partially transparent in order to be able to directly to oberseure the decoction pieas.

5.4 Inner pack materials shall boast certain characteristics of permeability, moisture-resistance, pressure-resistance and environmental friendliness, and shall be recycled or degraded.

5.5 Inner pack materials of decoction pieces shall be disposable and shall not be reused.

5.6 Decoction pieces administered as narcotic and toxic drugs shall comply with the relevant laws, regulations and standards of the current country.

5.7 Types of packing materials and the relevant scope of application are as follows:

5.7.1 Polyethylene plastic film shall comply with ASTM D3981 and ASTM D4635. It is suitable for decoction pieces that are prone to mold and insect damage.

5.7.2 Polyethylene blown film for packaging applications shall comply with ASTM D4635 and ASTM D4635,including 4 types: paperboard composite film (paper/plastic);fiber composite film (fiber/plastic);multi-layer composite film (plastic /plastic);aluminum foil composite film (metal/plastic). It is suitable for decoction pieces that are prone to midew and insect infestation.

6 Setting of Packing Dose Specifications

6.1 Setting Principles

Setting inner pack specifications of decoction pieces shall follow the principle "meeting the requirements of commonly used clinical dose, the least dose and expensiveness for high-frequency".

6.2 Specification Requirements

6.2.1 There are 13 inner pack specifications for decoction pieces: 1g, 3g, 5g, 6g, 9g, 10g, 12g, 15g, 30g, 250g, 500g, 1000g , >1000g.

6.2.2 Deviation of packaging amount of inner pack decoction pieces shall be consistent with the requirements of Table 1.

Table 1 Requirements on Packing Volume Deviation of Decoction Pieces with Inner Pack

Marked Weight or Packing Volume	Permitted Deviation from the Marked Weight or Packing Volume
1g	±10%
3g、5g	±6%
6g、9g、10g	±5%
12g、15g、30g	±4%
250g、500g、1000g、above 10000g	No less than 97% of the marked weight or packing Volume

Note 1: Other special requirements can be made for precious and scarce decoction pieces.

Note 2: Test method of packing volume deviation for the specifications of 1~30g is taking 10 test packages and weighing the weight of the content of each package respectively (the weight weighted shall be with an accuracy of one percent of the weight taken) and then comparing the packing volume of each package to the marked packing volume. No more than 2 packages can exceed the permitted packing volume variation and none of them can surpass twice of the permitted variation.

Note 3: Test method of packing volume deviation for the specifications of 250g or moreis is taking 3 test packages and weighing them respectively (the weight weighted shall be with an accuracy of one percent of the weight taken) and calculating the packing volume of each package and the average packing volume both of which shall be in accordance with relevant requirements of the above table. If one of them does not satisfy the requirement, then another three test samples will be checked and all the three of them should conform to the requirements.

6.3 Other Requirements

6.3.1 For drugs with anesthetic functions, such as pericarpium papaveris, the inner pack requirements are strictly in accordance with the regulations of relevant countries and regions.

6.3.2 For decoction pieces under the control of anesthetic and toxic drugs and marked as toxic in the *Pharmacopoeia of the People's Republic of China* 2020 edition, the inner pack quantity of decoction pieces shall not exceed the maximum dose stipulated by the relevant country or region.

6.3.3 For decoction piece whose dosage unit is not considered in terms of weight, such as scolopendra (one), there may not be inner pack specifications. Such decoction pieces shall be processed as per the dosage stated in the prescription.

7 Requirements for Labels

7.1 A label should be printed on or marked on the inner pack of decoction pieces.

7.2 The letters or characters on the label shall be clear and easy to recognize. The expression shall be scientific, standardized and accurate.

7.3 The marking shall be clear and striking. There shall be no faded lettering and no insecure pasting. Any modification or addition shall not be made by means of pasting, cutting or altering.

7.4 It is prohibited to use unregistered trademarks in the label, and it is prohibited to use the name of decoction pieces that have not been approved by the food and drug administration of the relevant country or region.

7.5 The labels of decoction pieces and toxic decoction pieces under the control of narcotic drugs shall comply with the relevant provisions of the relevant countries or regions.

7.6 For drugs with special requirements on storage, the requirements shall be marked on the notable place of the label.

7.7 No words, audio/visual or other materials unapproved be secretly inserted into the drug package to introduce or publicize the product or the enterprise.

7.8 The label of decoction pieces shall at least contain the following information:

a) Trademark;

b) Name (the standardized names listed in ISO 18668-2);

c) Place of production;

d) Packing volume;

e) Barcode (including QR code) (pursuant to ISO 18668-1);

f) Product batch number;

g) Date of manufacture;

h) Manufacturing enterprise;

i) The quality standard implemented.

8 Requirements for Quick Response (QR) Codes

8.1 QR Code Area

Under the premise of the realization of fast recognition, QR code area shall be bigger than one sixteenth of the area of the packing material. The QR code shall be attached to the packing material with adhesive stickers and labels or be directly printed on the packing material.

8.2 QR Code Scanning Results

QR code scanning results should at least include the following information:

a) Enterprise name;

b) Product name;

c) Product bar code;

d) Product category;

e) Chinese medicines codes (based on ISO 18668-1 and ISO 18668-2);

f) Brand;

g) Specification/type;

h) Place of production;

i) Storage methods (base on ISO 22217);

j) Net content;

k) Executive standards (*Pharmacopoeia of the People's Republic of China* 2020 edition or relevant standards);

l) Description of products (effect);

m) Enterprise website;

n) Manufacture factory;

o) Production address;

p) Period of validity for records.

9 Requirements for Colour Codes

9.1 The application of colour codes should follow the principle of markedness and obvious chromatic aberration. Different varieties of the same specification should use the same colour.

9.2 The colour code of the inner pack shall comply with the relevant provisions of the relevant country or region. The printed content, colour and luster shall meet the requirements.

9.3 The colour code shall be attached to the packing bag or label and marked on the central area of the packing material in the form of a colour stripe. The area of the colour strip should be bigger than one fourth of the area of the packing material.

9.4 The colour code has adopted the internationally Pantone colour card. Specifications of inner pack of decoction pieces and their corresponding colour code see Annex A.

9.5 Colour codes adopt the following printing methods:

a) Print different colours directly on the inner pack. This method is applicable to inner pack that uses colour stripes printing labels and directly prints dedicated labels on transparent plastic bags.

b) Use self-adhesive papers with several different underpaintings to print labels. Each colour represents a specification. This method is applicable to inner pack with labels attached such as fully transparent plastic bags.

Annex A
(Informative)
Inner Packaging Specifications for Decoction Pieces Corresponding Pantone Color Chart

The corresponding color codes of the inner packaging specifications of medicine pieces and Pantone color cards are shown in Table A.1

Table A.1 Inner Packaging Specifications for Decoction Pieces Corresponding Pantone Color Chart

Colour Card No.	Colour Used	RGB	HEX/HTML	Entity Sample	Specification
P 57-8 C	Red	211 22 59	D3163B		1g
16-1356 TCX	Persimmon Orange	244 115 39	F47327		3g
299 C	Blue	0 163 224	00 A3E0		5g
20-0084 TPM	Frosted Birch	209 198 185	D1C6B9		6g
1535 C	Brown	148 69 11	94450 B		9g
355 C	Green	0 150 57	009639		10g
2597 U	Purple	130 90 164	825 AA4		12g
14-0760 TCX	Yellow	255 212 0	FFD400		15g
8201 C	Light Steel Blue	103 137 156	67899 C		30g

Note : According to the actual clinical situation, specifications above 30g do not need to be color-coded.

Bibliography

[1] China Pharmacopoeia Commission. Pharmacopoeia of the People's Republic of China. English ed［M］. Beijing: China Medical and Technology Press, 2022.

[2] China Pharmacopoeia Commission. Pharmacopoeia of the People's Republic of China. Part 1［M］. Beijing: China Medical and Technology Press, 2022.

[3] Standardization Administration of China. Packaging Termas-Part 1-Basic： GB/T 4122.1-2008［S］. Beijing： Standards Press of China, 2011.

[4] National Technical Committee 48 on Plastics Products Standardization of China. Polyethylene Blown Film for Packaging Application： GB/T 4456-2008［S］.Beijing:Standards Press of China,2011.

[5] State Pharmaceutical Administration.Composite Film for Packing Applications–General Rules： YY 0236-1996［S］. Beijing: Standards Press of China, 1998.

[6] Liao L P, Lv A P, Zeng Q M. Experience on National Traditional Chinese Medicine (TCM) Standard Development- Interpretation of the Standards on Coding System of Chinese Medicines［M］. Beijing: Standars Press of China, 2016.

[7] Wu P K, Xu M Q. Handbook of Codes and Functions for Chinese Medicines and Formulas［M］. Beijing: Standars Press of China, 2016.